DRAWING GEOMETRY

Also by Jon Allen

Making Geometry: Exploring Three-Dimensional Forms

DRAWING GEOMETRY

A Primer of Basic Forms
for Artists, Designers
and Architects

Compiled and Drawn by Jon Allen

With a Foreword by Keith Critchlow

First published in 2007 by Floris Books
© 2007 Jon Allen
Foreword © 2007 Keith Critchlow
Eighth printing 2022

Jon Allen has asserted his right under the
Copyright, Designs and Patents Act 1988
to be identified as the Author of this Work

All rights reserved. No part of this publication may
be reproduced without the prior permission of
Floris Books, Edinburgh
www.florisbooks.co.uk

British Library CIP Data available
ISBN 978-086315-608-3
Printed in Great Britain
by Bell & Bain Ltd

Printed on sustainably
sourced FSC® paper.
Uses plant-based inks
which reduces chemical
emissions.

Contents

Acknowledgments 7
Foreword by Keith Critchlow 8
1. Introduction 10
2. Getting Started 12
3. Triangle: Three-Sided Figure 14
4. Square: Four-Sided Figure 16
5. Pentagon: Five-Sided Figure 18
6. Hexagon: Six-Sided Figure 28
7. Heptagon: Seven-Sided Figure 30
8. Octagon: Eight-Sided Figure 36
9. Enneagon: Nine-Sided Figure 40
10. Ten, Eleven, Twelve and Thirteen-Sided Figures 44
11. Dividing a Line 50
12. Miscellany 56
Appendix 64
Afterword: Explorers in Geometry 85
Further Study 86

Acknowledgments

There is little in this book that is original, unless one is using the word in its deeper sense of 'pertaining to origins' — in which case hopefully all of it is!

There is little in this book, too, that is not the fruit of my twenty plus years' collaboration with Professor Keith Critchlow, artist, teacher, master of geometry — and one of the century's 'rare conceptual minds.'* I therefore acknowledge an enormous debt to him as teacher and colleague.

In many ways this book is a collection of material that Keith has shared with me over the years as I learnt the practice of geometry; and which I now share with you in the hope that you may be equally inspired.

Any faults in the book are entirely my own, as are the accuracy calculations in the Appendix, carried out with the aid of a certain facility in algebra and trigonometry (for which I have my old grammar school to thank).**

Warm thanks are due to my publisher, Floris Books, and in particular to Christopher Moore.

Deep gratitude also to my dear wife Clare.

---

* According to Buckminster Fuller who, being one himself, was well placed to recognize another.

** Where vestiges of a classical ('Liberal') education were to be found in the class names: Figures, Rudiments, Grammar, Elements, Syntax, Poetry, and Rhetoric.

Foreword

Geometry is literally 'earth (geo-) measure,' according to its Greek language origins. In order to elevate this objective subject, Plato suggested that 'stereometry' — or 'space measure' — was a better word to use, being more abstract, universal and nearer to the source of all objectivity.

Kairos educational charity, in its objective of integrating the traditional arts and sciences, has been a focus of propagating sacred and philosophical geometry since it was founded by myself in 1984. Since then, it has generated and inspired much activity and personal revelations to many students and practitioners. Jon Allen has gathered here a collection of these various 'discoveries.' It is not possible to attribute every construction, and therefore we are providing by way of acknowledgment a list of participants from over the years, some being students of mine on the VITA (Visual Islamic and Traditional Arts) programme at The Prince's School of Traditional Arts. I encouraged Jon, as well, to check the accuracy of the constructions by algebraic means in order to affirm the depth of objectivity, and these calculations are included in an Appendix. However, the initiative for this book is solely Jon's own.

Geometry is primarily a philosophical and therefore, to many, a sacred art/science. It was one of the four obligatory educational studies for those who have contributed most to the great civilizations of the human family. The other three subjects, or disciplines of universal objectivity are: first, Arithmetic founded on pure number; second, Music founded on number in time; and third, Astronomy (synonymous with astrology to the ancient sages), being number in time *and* space. Thus 'Number' is the most important of the objective studies, considered by Iamblicus — who represented the Pythagoreans — to be the means by which the human mind could become closest to the Divine Mind or the First Principles of Intelligibility.

It is not widely known that the thinkers of the ancient world — among them Pythagoras, Plato, Aristotle and Euclid — did not have written numbers to work with. For them, numbers were represented by alphabetical letters *or,* most importantly, by round pebbles *(khalix,* which became in Latin *calcis,* and the root of our current word 'calculation'). When these pebbles became strung in lines, we gained the abacus.

These round pebbles are the integral source of Geometry *and* Arithmetic together — as one pebble represented the 'point' in Geometry; two represented the geometrical line (a point at each end); three became the natural representation of the triangular plane (when touching); and finally, four became the first solid figure as the Tetrahedron. This integral approach to the meaning and morphic significance of 'point,' 'line,' 'plane,' and 'solid' was the basis of my first publication, *Order in Space.*

Geometry is thus at least a fourfold subject: first, given its source in pure universal objectivity (traditionally, in the unchanging Divine Mind of the archetypes); second, as the philosophical and *actual* expression supporting each of the Revelations accordingly; third, as a discipline to aid the development of the soul of each participant; and finally, in its practical use as a craft discipline for those who make material objects for this world. In this current volume, Jon Allen, with his excellent illustrations, has offered a method to the 'makers' of our time, that is, Artists, Designers and Architects. We at Kairos hope that it will lead those who make a living out of Geometry into the deeper realms of its timeless nature, a nature which touches upon the mystery of how we may move from the imprisonment of time into the freedom of eternity. It is only here that we may escape the tyranny of uncertainty and find our true home in certainty. As an anthology of old and new methods, this is an excellent starting point. We hope it will lead the reader into the deeper mysteries of meaning, and thereby uplift the soul.

Keith Critchlow
March 2007

1. Introduction

This is a book of practical instruction in drawing two-dimensional geometry by hand. The primary regular polygons constructed in this book (triangle, square, pentagon, and so forth) are the foundation of geometry.* That they are all contained within a circle (symbol of unity) can be taken as a profound truth.

This book does not cover three-dimensional geometry, or pattern making. Nor does it consider the use of geometry and proportion in architectural design, its relation to number symbolism, or its place in philosophy and cosmology. These are all vast subjects wherein geometry plays a major role, but beyond the scope of this small book. Nevertheless, they exist as a field of universal resonance accompanying all geometry, which by its very nature is objective (not the product of personal imagination) and unchanging over time; thus qualifying for description in the Platonic tradition as 'the ever-true.'

All of the diagrams in this book are hand drawn. There is great value in drawing by hand, and good reason to resist the temptation to resort to a computer. We lose something when we use computers to draw geometry, and however beguiling their mechanical precision, they lack 'heart': in some subtle way we become observers rather than participants. An activity worth doing is worth doing well, and as much is gained in the doing as in the final product. To surrender the experience of drawing by hand for the convenience of digital storage, revision and transmission seems a great shame — and essentially misses the point that drawing geometry is as powerful a meditation, or as inspiring a creative activity as it is possible to find. Look around in nature, at the history of art and architecture of the world, or in science and you will find geometry hand in hand with Truth and Beauty.

The constructions in this book are 100% accurate, unless otherwise noted. The rest are very close approximations (often 99% or so).** The actual exactness of your own constructions will also depend on the precision of your drawing, which will get better with practice and attention.

To see how keen your attention is — and how sharp your pencil! — try the construction of a hexagon (Figures 6.1 and 6.2) which is also a useful 'warm up' exercise at the start of any geometry session. Did you return exactly to the starting point after swinging six radii around the circumference, or were you just a little off? If so, because we know this construction is perfectly exact, any inaccuracy is in your drawing technique. Check that your pencil is sharp and, being careful to place your compass point exactly (using your second hand as a steadier), try again. You will usually find that with a bit of practice,

---

* 'Polygon' simply means many-sided (literally 'angled'). A 'regular' polygon is one in which all sides (and angles) are equal.

** The mathematical calculations, and a brief note on the question of accuracy, are included in the Appendix.

your technique will sharpen up nicely. This exercise will also show you the level of care required to get the more complex constructions to work as they should.

It is not necessary that you work through the book from beginning to end in order. It is intended that you can dip in and draw any construction on its own. However, if you are new to drawing geometry, you might find it useful to try the earlier constructions first as they are the simplest, and will introduce you to methods used throughout the book.

If you are new to geometry, I hope you find the instructions clear, and are inspired by the drawings you produce. If you are an old hand, and simply want to use this book from time to time to remind yourself of a method or two, I hope that the book gives you what you need — and a little more to ponder.

If this book proves not only useful and enjoyable, but also opens your eyes to the geometry that is all around, then the author will be satisfied that this volume has served its purpose.

2. Getting Started

You will need the following equipment (see photograph: *Tools of the Trade*):

- a good pair of compasses (with screw wheel adjuster is best), and a pen attachment if you want to draw in ink or coloured pencils. It's useful but not essential to have several different sizes of compass, each of which can be set and kept at different radii;

- pencils (different coloured leads can be useful) and drafting pens (if you want to make an ink drawing);

- a straight edge (ruler or set square);

- some paper (tracing paper for ink drawings; plain cartridge for pencil; watercolour or handmade paper if you intend to make a painting);

- a board or surface (which doesn't mind repeated piercing by your compass point);

- a place to work with good (ideally) natural lighting.

Tools of the Trade
30:60 set square (on the left), an adjustable setsquare (on the right); a pair of compasses (below) with pen attachment (above), with a small pair of compasses (beside); an assortment of pencils and a technical drawing pen (at the top); and (to the left) a special small circle gadget, which I find better than a compass for marking points.

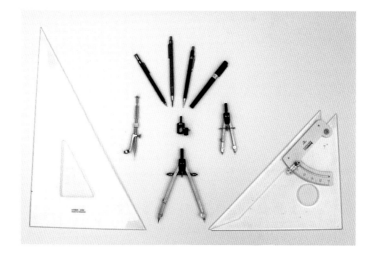

Useful tips

⁂ keep your leads sharp;

⁂ guide your compass point into place with your second hand: the more precisely you place the compass point, the more accurate your drawing will be;

⁂ periodically check that your compass hasn't accidentally adjusted itself;

⁂ before starting your drawing, check that the full construction will fit on your page (some constructions begin in the centre of the construction; but others to one side). Check also that your compasses will stretch to the widest arc required in the construction (usually the *diameter* of the first circle);

⁂ before drawing in ink, I recommend that you first draft in pencil.

Other ways to draw

It is possible, and a good deal of fun, to draw geometry at a larger scale, for example on a beach (first find the right kind of sand) or a hard-surfaced playground. For this you will need some friends, and slightly different equipment: for the beach: steel pegs* and mallet and string of different colours; for the playground: steel pegs, string, chalk and a mason's chalked line.

The author and friends drawing geometry on a baseball court

---

* To hold one end of the string when drawing circles and arcs. But don't hammer into the surface without permission!

3. Triangle: Three-Sided Figure

How to draw an equilateral triangle in a circle

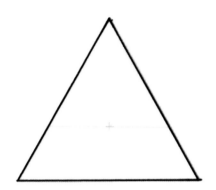

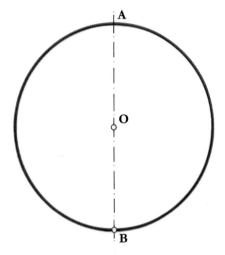

Figure 3.1. Step 1
Draw a vertical line. On this line, draw a circle with centre O *(shown coloured)* cutting the line at points A and B.*

---

* Or we could call this 'circle O.'

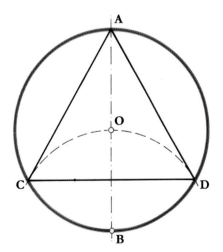

Figure 3.2. Step 2
With compass point at B, and the same radius as the given circle, draw an arc passing through O and cutting the circle at C and D. Connect points A, C and D. ACD is an equilateral triangle.

3. TRIANGLE: THREE-SIDED FIGURE

How to draw an equilateral triangle on a line

Figure 3.3.
On any given line AB *(shown coloured)*, set compass radius to AB, and with compass point first on A and then on B draw arcs to intersect at C. Draw lines to connect A, B and C. ABC is an equilateral triangle.*

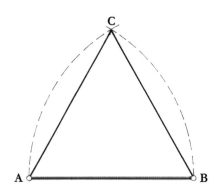

Figure 3.4.
By swinging the arcs below as well as above the line, we draw a shape that has come to be called the *Vesica Piscis.***

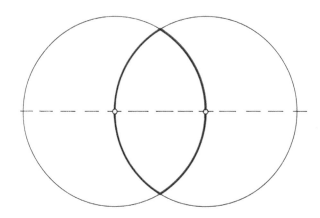

* This is the basis of the First Proposition in Euclid's *Elements*.

** The vesica shape has profound significance within Christian art, and indeed as a universal symbol of the 'womb of Creation.'

15

4. Square: Four-Sided Figure

How to draw a square in a circle

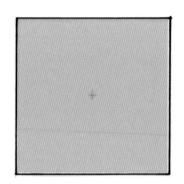

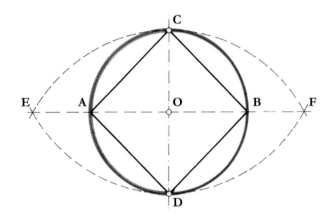

Figure 4.1.
On given circle centre O *(shown coloured)*, draw a vertical diameter CD passing through centre O. With compass point at C and radius CD, draw arc EF passing through D. With compass point at D draw arc EF with the same radius passing through C. Draw a line to connect points E and F (which line is at right angles to CD), cutting the circle at points A and B. ACBD is a square — in this diagram shown in the 'dynamic' position (see Critchlow 1976), which most people perceive as a 'diamond.'

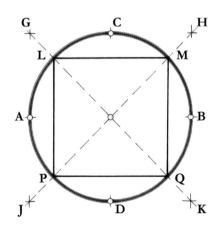

Figure 4.2.
On the same diagram, with compass radius set at that of the original circle, draw arcs from points A, C, B and D intersecting at new points G, H, J and K. Join G to K, and J to H, crossing the circle at L, M, Q and P. Join these new points to produce the 'static' square LMQP.

4. SQUARE: FOUR-SIDED FIGURE

How to draw a square on a line

Figure 4.3. Step 1
On any given line AB *(shown coloured)*, with compass point first on A and then on B, draw two circles with radius AB. Extend AB to cut the circles at E and F.

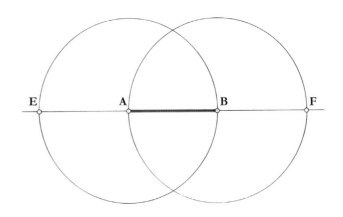

Figure 4.4. Step 2
From points E and B, then F and A, with a radius approximately three quarters of the circle diameter EB, draw arcs above and below each circle. These arcs will intersect at points G, H, J and K. Join G to J, and H to K, cutting the circles at points D and C. ABCD is a square, as is ABML.

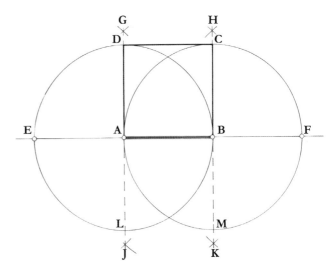

5. Pentagon: Five-Sided Figure

How to draw a pentagon in a circle

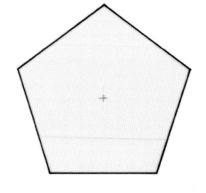

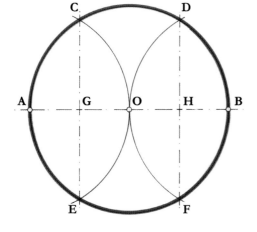

Method 1

Figure 5.1. Step 1
Draw a circle with centre O. Draw a diameter AB on the horizontal. With radius the same as the original circle, and centres A then B, draw arcs EC and FD passing through O. Connect CE and DF, cutting AB into four at points G and H.

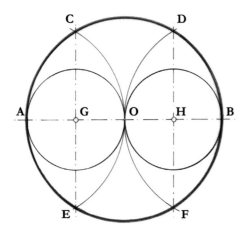

Figure 5.2. Step 2
Draw two circles centred on G and H, radius GO. The two smaller circles are the 'octave' of the original circle. They also give the familiar (Ying:Yang) symbol of interwoven duality, illustrated here:

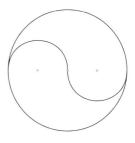

Figure 5.3. Step 3
With radius AH or larger, draw arcs from points A and then B above and below the original circle, intersecting at J and K. Connect J and K to create the vertical axis passing through O and cutting the circle at L. From L draw an arc to tangent both small circles G and H, cutting the main circle at points M and N. MN is the base of a pentagon.

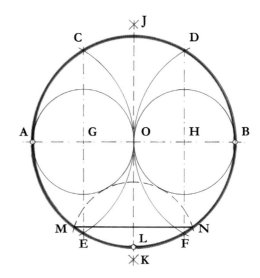

Figure 5.4. Step 4
With radius MN draw arcs from N to find P, and from M to find R — both points on the original circle. With the same radius draw arcs from P and R to cut the original circle at point Q on the vertical axis (JK). MNPQR is a pentagon.

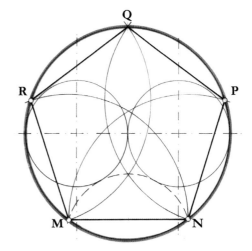

5. PENTAGON: FIVE-SIDED FIGURE

How to draw a pentagon in a circle

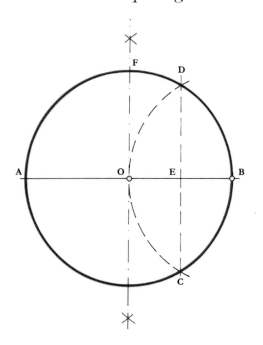

Method 2

Figure 5.5. Step 1
Draw a circle with centre O and radius OA. Draw the horizontal diameter AB. With the same radius and the compass centred on B, draw an arc to cut the circle at C and D. Join C to D to give E, which bisects OB. Then using points A and B and a convenient radius, draw arcs to establish the vertical axis passing through O and point F on the circle.

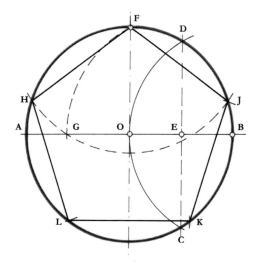

Figure 5.6. Step 2
With your compass centred on E and radius EF, drop an arc down to the horizontal axis, meeting it at point G. Then with compass on F and radius FG, swing an arc to meet the circle at H and J. From these two new points, and keeping the same radius, find points K and L. FJKLH is a pentagon.

5. PENTAGON: FIVE-SIDED FIGURE

How to draw a pentagon within a vesica

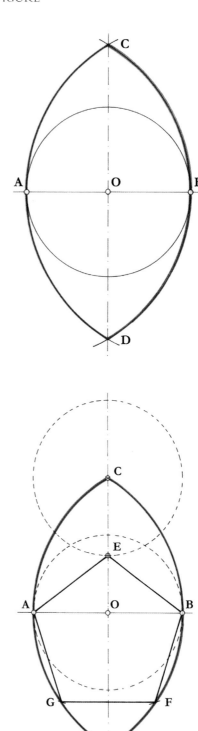

Figure 5.7. Step 1
Draw a circle with centre O. Draw the horizontal diameter AB, and then construct a vesica by drawing arcs with radius AB, and centres first A then B. Draw the vertical axis from the points C and D where the arcs of the vesica intersect, passing through O.

Figure 5.8. Step 2
With your compass set to the radius of the original circle, and centred on point C, draw a circle. This intersects the vertical axis at E, the apex of our pentagon. With compass on point E, and radius AB, swing your compass to cross the vesica at points F and G. EBFGA is a pentagon to an accuracy of 99.25%. (Construction by Paul Marchant.)

5. PENTAGON: FIVE-SIDED FIGURE

How to draw a pentagon from a hexagon

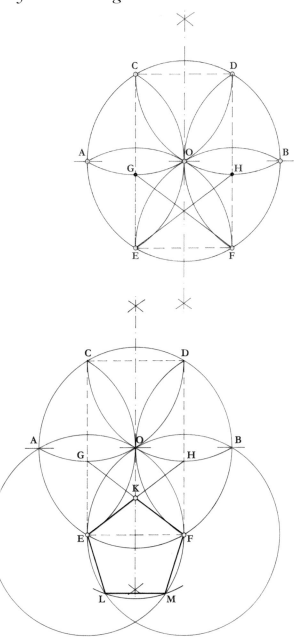

Figure 5.9.
First draw the *Flower of Life* pattern (see Figures 6.1 and 6.2). By placing your compass in turn on points A and then B, and with a convenient radius (longer than AB), draw arcs to intersect above and below the circle, and join these intersection points to draw the vertical axis. Then by joining points C, D, F and E draw the rectangle CDFE (which is a √3 rectangle). Where the sides of this rectangle cross petals A and B mark points G and H. Join H to E, and G to F.

Figure 5.10.
Lines EH and GF cross on the vertical axis at point K, which is the apex of a pentagon with two sides EK and KF. With your compass point on F, and radius that of the original circle, complete the circle that earlier gave us arc EOB. Similarly, complete the circle E which continues the arc AOF. With the same radius, and compass point on K, draw an arc to cross these new circles; which gives us the final two points of the pentagon, L and M. This construction is 99.4% accurate.

5. PENTAGON: FIVE-SIDED FIGURE

How to draw a pentagon on a line

Method 1

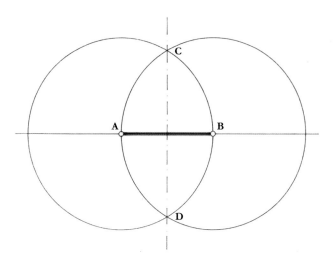

Figure 5.11. Step 1
Let AB be the given line. Draw two circles with centres A and B and radius AB. Extend line AB to meet both circles. Draw the vertical axis by joining the intersection points C and D.

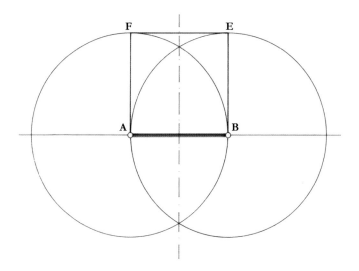

Figure 5.12. Step 2
Construct a square ABEF on AB (compare with Figure 4.4).

23

5. PENTAGON: FIVE-SIDED FIGURE

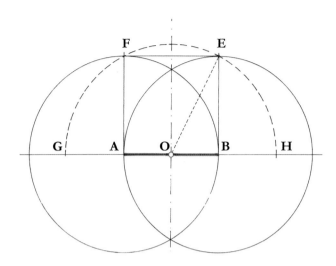

Figure 5.13. Step 3
With compass point on O and radius OE (the diagonal of half a square) draw an arc through E and F extended to meet the horizontal axis at points G and H.

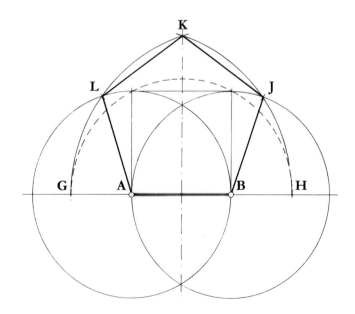

Figure 5.14. Step 4
With compass point on A and radius AH, draw an arc upwards to cut circle B at J and the vertical axis at K. Similarly, with compass on B and radius BG (which is the same as AH), draw an arc up to meet point K, cutting circle A at L. Join up points A, B, J, K and L. ABJKL is a pentagon.

5. PENTAGON: FIVE-SIDED FIGURE

How to draw a pentagon on a line

Method 2*

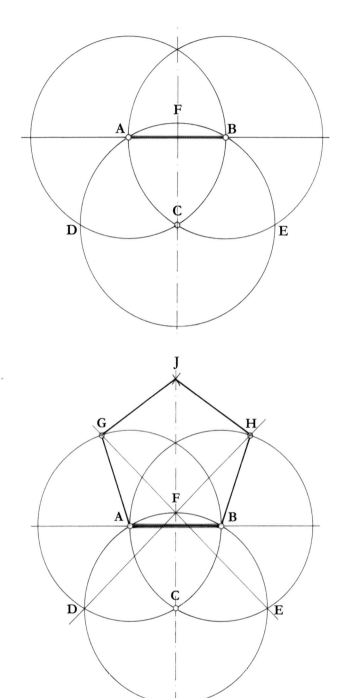

Figure 5.15. Step 1
On line AB, draw two circles with radius AB, centred on A and then B. Using the crossing points of these circles, draw the vertical axis. On point C draw a third circle with the same radius: this will cross the vertical axis at point F and the first two circles at points D and E.

Figure 5.16. Step 2
From D draw a line through F to cut circle B at H. Similarly, from E draw a line through F to find G on circle A. From points H and G, and with radius AB, draw arcs to intersect on the vertical axis at J. ABHJG is a pentagon. This construction is 99.66% accurate.

---

* This construction was first revealed and used by Albrecht Dürer; and is referred to by John Michell (1973) p.75.

5. PENTAGON: FIVE-SIDED FIGURE

Improving accuracy

The accuracy of the above construction (Figure 5.16) can be improved by making one further step as follows:

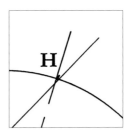

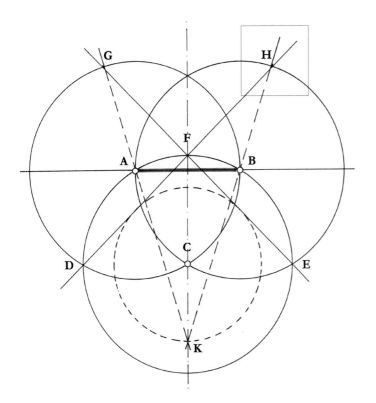

Figure 5.17.
With the compass centred at C, draw a circle to tangent lines DH and EG, cutting the vertical axis at point K. Draw a line from K through B to cross circle B close to point H. Similarly, draw a line from K through A to cut circle A close to point G. These two new points give the pentagonal positions to an accuracy of 97.96% — but to the other side of the true position compared to our previous construction (Figure 5.16). Therefore, even more exact points for the pentagon will lie between the two lines cutting circles B and A (see enlarged detail of point H above).

5. PENTAGON: FIVE-SIDED FIGURE

How to draw a pentagon on a line

Method 3

Figure 5.18.
On line AB draw circles centred on A and then B with radius AB. Draw a line through their intersection points C and D to establish the vertical axis and the centre point O. Extend AB beyond the circles, crossing at points E and F. With compass on point O and radius OE draw an arc to cross the vertical axis at G. Draw lines from G through A and B to intersect the two circles at H and K. With the original radius AB, draw arcs from H and K in turn to give point J. ABKJH is a pentagon to an accuracy of 97.6%.

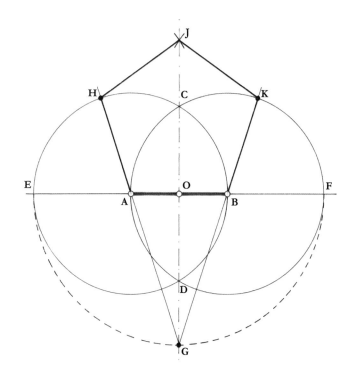

6. Hexagon: Six-Sided Figure

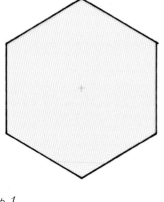

How to draw a hexagon in a given circle

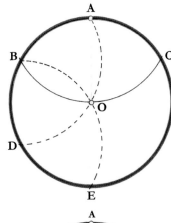

Figure 6.1. Step 1
Draw any circle with centre O and radius AO (A being any point on the circumference). With compass point on A and the same radius, draw an arc BC from one side of the circle to the other, passing through the centre O. From B draw an arc from A through O to new point D on the circle: similarly from point D draw an arc from B to find point E.

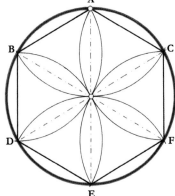

Figure 6.2. Step 2
Continue around the circle drawing arcs from points E, F and C until you have drawn the pattern above. ABDEFC is a hexagon, composed of six equilateral triangles (marked here with dotted lines). The overall pattern is known as the *Flower of Creation*. Although this is the most often mentioned, there are others similar. Here is a fuller set (from Kairos foundation):

 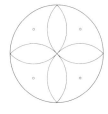 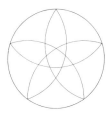 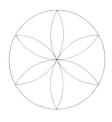 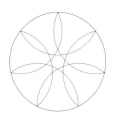

Flower of Spirit   *Flower of the Elements*   *Flower of Life*   *Flower of Creation*   *Flower of Intelligence*

6. HEXAGON: SIX-SIDED FIGURE

How to draw a hexagon on a line

Figure 6.3. Step 1
On any given line AB, with radius AB, draw two circles centred on A and then B. Where they intersect at point O, with the same radius, draw a circle passing through A and B, cutting the first two circles at points C and F.

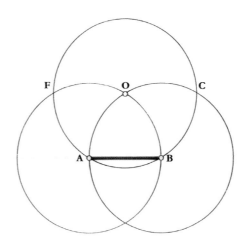

Figure 6.4. Step 2
Keeping the same radius, draw an arc from C to cut the first circle at D; and an arc from F to cut the second circle at E. ABCDEF is our hexagon.

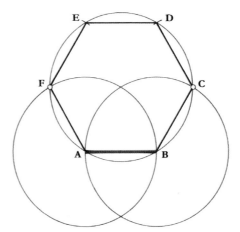

7. Heptagon: Seven-Sided Figure

How to draw a heptagon in a circle

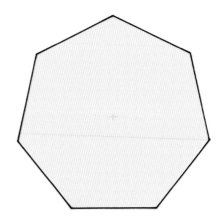

Method 1

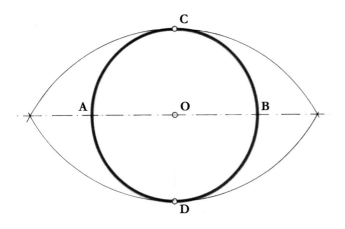

Figure 7.1. Step 1
Draw any circle with centre O. Draw a vertical diameter CD. From point C with radius CD draw an arc to tangent the circle at point D. From D draw a similar arc through C. Join the points where the arcs intersect to create the horizontal axis passing through the circle at points A and B.

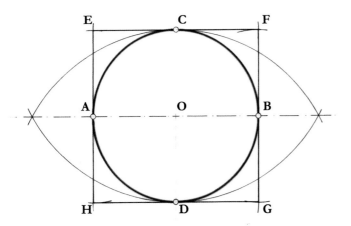

Figure 7.2. Step 2
With radius AO, draw arcs from points A, C, B and D to intersect at corner points E, F, G and H. EFGH is a square.

7. HEPTAGON: SEVEN-SIDED FIGURE

Figure 7.3. Step 3

From points E and F draw two arcs with radius EF, intersecting at point J within the square. EFJ is an equilateral triangle, which crosses our first circle at points K and L. CK and CL are two sides of a heptagon to an accuracy of 99.9%.

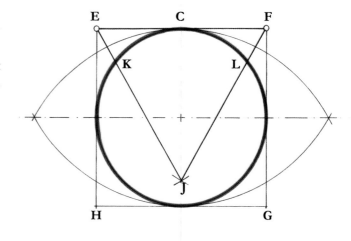

Figure 7.4. Step 4

Setting your compass radius to CK, proceed around the circle from point K to mark the other four points of the heptagon within the circle. Proceed round to point L to check the accuracy of your construction, and adjust as necessary. CKMNPQL is our heptagon.* (Construction by John Michell.)

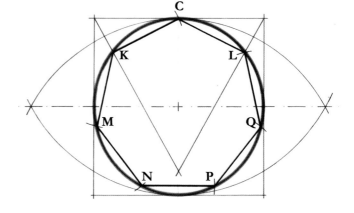

---

* The use in this construction of an equilateral triangle and a square very nicely demonstrates 3+4=7.

7. HEPTAGON: SEVEN-SIDED FIGURE

How to draw a heptagon in a circle

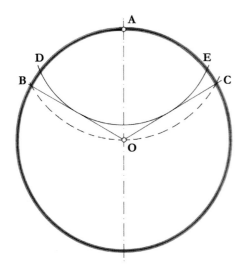

Method 2

Figure 7.5. Step 1
Draw a circle with centre O. Draw a vertical diameter crossing the top of the circle at A. Keeping the same radius, and placing the compass on A, swing an arc passing through O to cut the circle in two places, B and C. BA and AC are two sides of a hexagon within the circle (see Figure 6.1). Draw the radii OB and OC; and then with compass still centred on A draw an arc to tangent these two radii, which will cross the circle at points D and E. DA and AE are two sides on a heptagon.

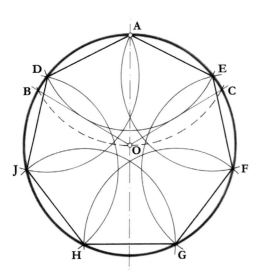

Figure 7.6. Step 2
To complete the construction of the heptagon, place the compass on E with radius AE and swing an arc to find point F on the circle; with compass on F and the same radius swing an arc to find G; and so on around the circle. If you do not arrive back at the starting point, you will need to adjust your compass slightly to produce an accurate heptagon. This construction is 99.79% accurate.

How to draw a heptagon in a circle

Method 3

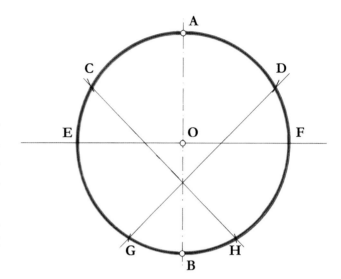

Figure 7.7. Step 1
Draw a circle with centre O. Draw a vertical diameter AB. From A, and with the same radius, mark the points C and D of the hexagon. Find the horizon by drawing arcs from points A and B with a larger radius (in this diagram a radius of AB is used); the horizon cuts the circle at E and F. With your compass set to radius EC* mark points G and H by setting your compass on B. Join C to H, and D to G.

---

* The chord of a twelve-sided figure.

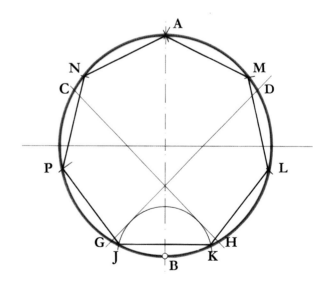

Figure 7.8. Step 2
With your compass point again on B, draw an arc to tangent GD and CH. This arc cuts the circle at points J and K. JK is one side of a heptagon to an accuracy of 99.26%. Find the rest of the figure by swinging arcs of radius JK from K, L, M and G, P, N in turn. If you do not arrive exactly at point A, adjust your compass slightly and walk round again from K on one side, and J on the other.

7. HEPTAGON: SEVEN-SIDED FIGURE

How to draw a heptagon on a line*

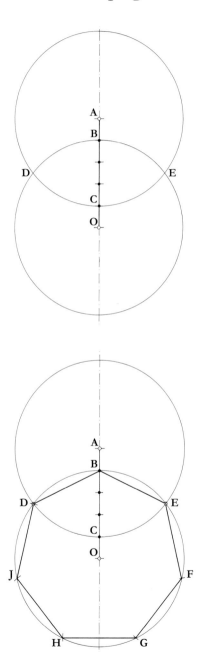

Figure 7.9. Step 1
Begin by drawing a vertical line, and on it mark AO measuring exactly 5 convenient units. With a radius of 4 such units, draw two circles, one centred on A and the other centred on O. The first will pass through C; the second through B — and the two circles will intersect at points D and E.

Figure 7.10. Step 2
DB and BE are the first two sides of a heptagon contained within circle O. The remaining points F, G, H and J can be found by setting your compass to a radius of EB, and 'walking round' the circle, marking arcs lightly as you go, and then working around again adjusting the radius slightly as necessary to arrive exactly at point D. This construction is 99.78% accurate.

---

* This construction is given in *Sacred Geometry* by Miranda Lundy.

How to draw a heptagon within a vesica

Figure 7.11. Step 1
Begin by constructing a vesica on the line AB by drawing arcs in turn from points A and then B with radius AB, intersecting at points C and D. Find the centre point O by joining C and D, crossing AB at its midpoint O. Now bisect OB by drawing arcs from O and then B with radius OB, and connecting the intersection points to give quarter point E. Now draw a circle centred at E, and with radius EO. With the same radius, and your compass on C, draw a circle which intersects the vertical axis at F.

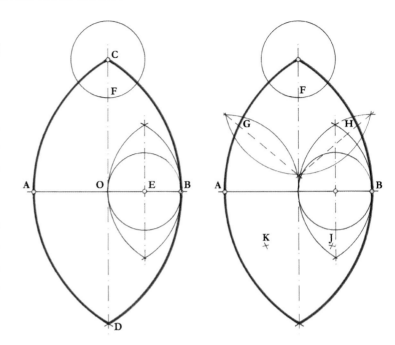

Figure 7.12. Step 2
Point F is the apex of our heptagon. To find the two adjacent points, which lie on the vesica, we need to bisect FB and FA. Do this by setting your compass in turn on points F, B and A, and drawing arcs of the same radius (in the diagram opposite, the radius used is OB). By drawing lines to connect the intersection points of the arcs, we can find points G and H on the vesica (and also, incidentally, we have found the centre point of the heptagon where these two lines cross on the vertical axis). We now have five points of our heptagon (A, G, F, H and B). To find the two remaining points, place your compass on G and then B with radius GH, and draw an arc below the line AB: this will give point K. Similarly, with compass on points H and A, and radius still GH (which is the same as AH), draw arcs to find point J.

Figure 7.13. Step 3
By joining up the points A, G, F, H, B, J, K we have our heptagon (to an accuracy of 98.25%). ▼

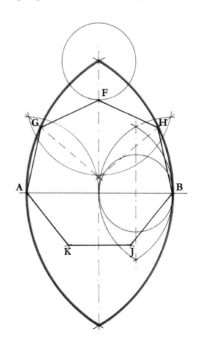

35

8. Octagon: Eight-Sided Figure

How to draw an octagon inside a circle

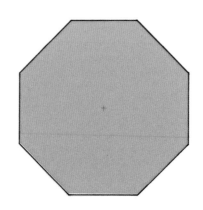

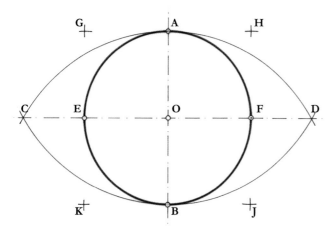

Figure 8.1. Step 1

Draw a circle centred at O and with radius OA. Draw the vertical diameter AB. With compass first on A, and then on B, and with radius AB, draw half circle arcs which intersect at points C and D. Draw the horizontal axis by connecting C to D, which cross the circle at E and F. With the original radius (OA), and compass in turn on A, F, B and E, draw arcs to intersect at corner points G, H, J and K.

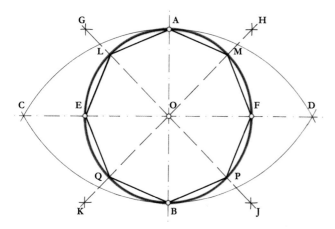

Figure 8.2. Step 2

Draw the diagonals by connecting point G to J, and H to K. These diagonals cut the circle at points L, M, P and Q. Draw the 'dynamic' octagon by connecting points A, M, F, P, B, Q, E and L.

8. OCTAGON: EIGHT-SIDED FIGURE

How to draw an octagon outside a circle

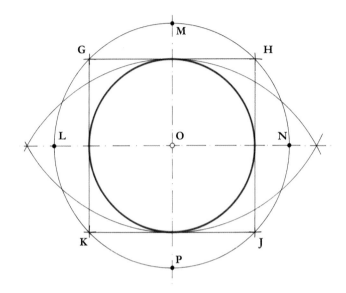

Figure 8.3. Step 1
Start by constructing figure 8.1. Then proceed to connect the four corner points G, H, J and K to create the square GHJK around the circle. Place your compass at the centre point O, set your radius to OG, and draw a circle around the square. This circle passes through the horizontal axis at points L and N; and the vertical axis at points M and P.

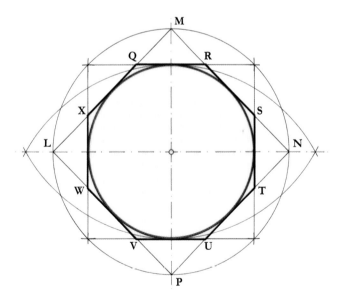

Figure 8.4. Step 2
Connect L to M; M to N; N to P; and P back to L. This 'dynamic' square cuts the 'static' square GHJK at the eight points of our octagon — QRSTUVWX. Note that this construction combines Figures 4.1 and 4.2, and can be seen as the balanced combination of the dynamic and the static aspects of the square. (See Critchlow 1976.)

8. OCTAGON: EIGHT-SIDED FIGURE

How to draw an octagon on a line

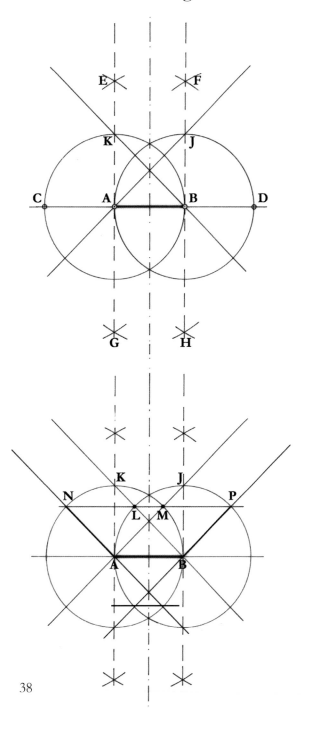

Figure 8.5. Step 1
Draw a horizontal line AB, which is to be one side of our octagon; and let the line extend some way to either side. With radius AB, draw two circles centred on A and B, crossing the horizontal at C and D. Draw the vertical axis by connecting the points where the circles intersect. With radius CB, draw arcs from C, A, B and D to intersect at points E, F, G and H, and connect E to G and F to H to draw lines perpendicular to A and B. Draw diagonal lines from A and B to pass through points J and K where these perpendiculars cross the circles: again let the lines extend either side.

Figure 8.6. Step 2
Draw a horizontal line through the two points L and M where the diagonals AJ and BK cross the inner arcs of the two circles; and extend this line to cross the outer arcs of the circles at points N and P. NA, AB and BP are three sides of our octagon.

8. OCTAGON: EIGHT-SIDED FIGURE

Figure 8.7. Step 3

With radius AB, draw two circles centred on N and P. These intersect the diagonals BK and AJ at points Q and R, giving us two further points of the octagon.

Figure 8.8. Step 4

Draw lines to join N to R, and Q to P, crossing the vertical axis at O, which is the centre of our octagon. Place your compass on O, set the radius to OA (checking first that it will pass exactly through N, Q, R, P and B) and draw a circle around the emerging octagon. Where this circle intersects the perpendiculars drawn through A and B (extend the lines if necessary) we find the final points S and T of our octagon.

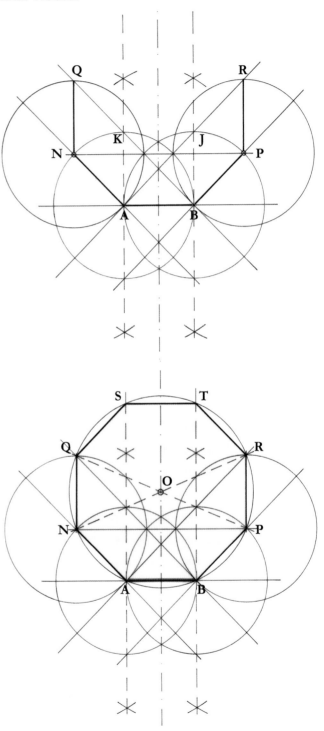

9. Enneagon: Nine-Sided Figure

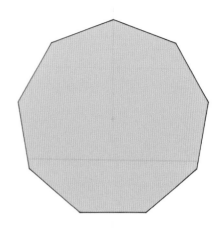

How to draw an enneagon in a circle

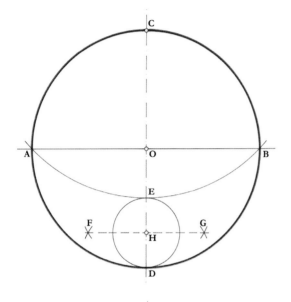

Figure 9.1. Step 1
Draw a horizontal line, and on it draw a circle with centre O, giving us A and B. Construct the vertical axis (as shown for example in Figure 4.1), giving us points C and D. Place compass point on C, and with a radius of CA, swing an arc from A to B, cutting the vertical axis at E. Bisect ED to find H. On H draw a circle with radius HE to fit between arc AB and the original circle.

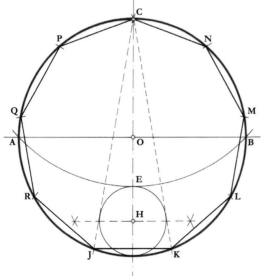

Figure 9.2. Step 2
From C draw lines to tangent this small circle on either side and intersect the original circle at points J and K. JK is one side of our enneagon. Using the length JK as a radius, walk round the original circle marking points L, M and N, arriving at point C. Likewise, travel round from J to TR, Q and P, arriving again at C. You will find that you fall slightly short of C, so adjust the compass by trial and error until you can walk round the circle exactly thirteen times. We now have our enneagon JKLMNCPQR. This construction is 95.8% accurate.

9. ENNEAGON: NINE-SIDED FIGURE

How to draw an enneagon in a vesica

Figure 9.3. Step 1
On a line AB draw the vesica CAD/CBD. Join CD to find the centre point O. Join A to C, and B to C, giving us an equilateral triangle ACB. Bisect AO and OB by drawing arcs with radius AO from A, O and B in turn to intersect at points E, F, G and H. Join E to F and G to H to find the quarter points J and K.

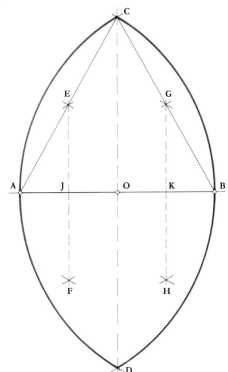

Figure 9.4. Step 2
Place your compass point on J, and with radius AJ draw a circle. Likewise, with your compass point on K, and with the same radius, draw a second circle. These circles tangent each other, and also the vesica. Join points J to C, and K to C, crossing the circles at points X and Y. Now place your compass point on C, and with radius CX, draw an arc to tangent the small circles. This arc will intersect the equilateral triangle at points L and M. We now have four points of our enneagon: A, L, M and B.

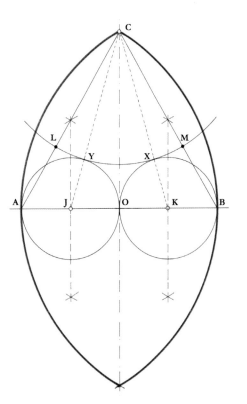

41

9. ENNEAGON: NINE-SIDED FIGURE

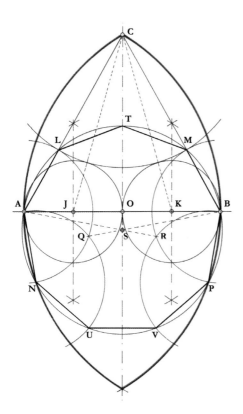

Figure 9.5. Step 3
Find the remaining points of the enneagon as follows: with compass radius set to AL, draw an arc centred on A from L round to N on the vesica; and centred on N round from A to the vesica. Similarly, draw arcs centred on B and then P. Draw a line from A to the intersection points of the arcs at R: and from B to the opposite intersection point Q. These lines cross the vertical axis at S. With compass on S and radius SA, draw a circle: this circle will encompass the enneagon. Where this new circle crosses the vertical axis we find the apex of the figure at T; and where the circle crosses the previous arcs from N and P we find the two lower points U and V. This construction is 99.6% accurate.

Symmetry in nature

Three, four, five and six-fold symmetries are very commonly found in nature. Other symmetries less so — look out for the infinite variety!

10. Ten, Eleven, Twelve and Thirteen-Sided Figures

How to draw a decagon (ten-sided figure)

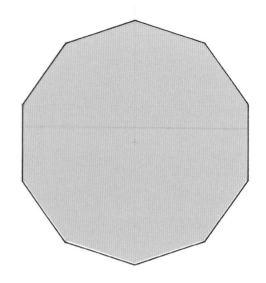

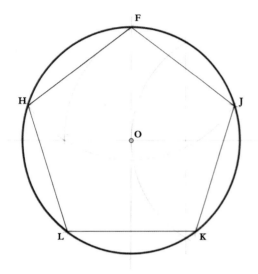

Figure 10.1. Step 1
Begin by drawing a pentagon (in this case following Method 2, Figures 5.5 and 5.6).

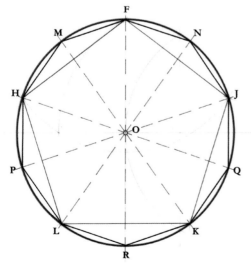

Figure 10.2. Step 2
Draw a diameter from each point of the pentagon, through the circle centre O, and out to the other side of the circle. This gives us points midway between the five points of the pentagon, and hence our ten-sided decagon FNJQKRLPHM.

10. TEN, ELEVEN, TWELVE AND THIRTEEN-SIDED FIGURES

How to draw an eleven-sided figure

Figure 10.3. Step 1
Proceed as for dividing a line into seven (Figure 11.6), but arrange your construction (i.e. AB) vertically.

Figure 10.4. Step 2
Using the seventh division as a radius, draw four circles as shown centred on points x, y, z and B. The uppermost circle tangents the parent circle at A, and the lowest circle cuts the parent circle at points L and M. LM gives us the side length of an eleven-sided figure to an accuracy of 99.6%. Complete the figure by using the length LM and walking your compass round the circle, adjusting until the circle is divided into eleven.*

---

* The relationship between 11 and 7 in this construction mirrors the traditional approximation to π of 22/7.

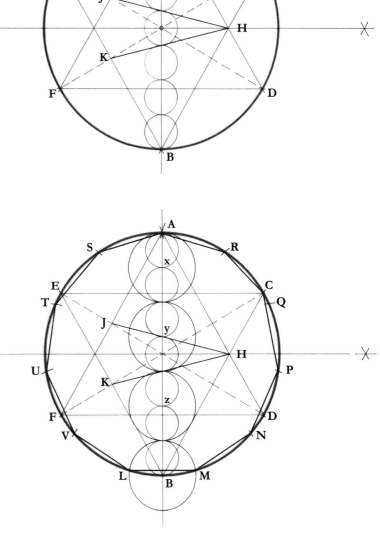

45

How to draw a dodecagon (twelve-sided figure)

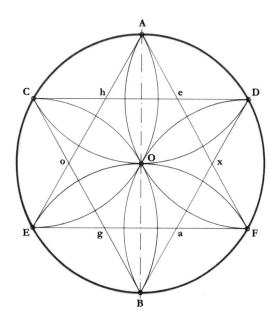

Figure 10.5. Step 1
Proceed as for a hexagon (see Figures 6.1 and 6.2) except that instead of drawing the hexagon ADFBEC, draw the hexagram (star) by connecting C to D and B; A to E and F. This gives us the crossing points h, e, x, a, g and o.

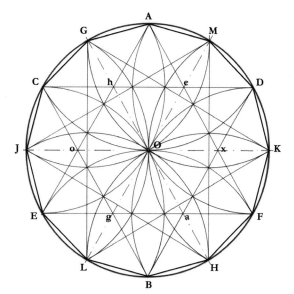

Figure 10.6. Step 2
Draw lines connecting h and a, passing through the circle centre O, and projected out to cross the circle on both sides: giving us points G and H. Similarly, connect e and g, projected to the circle to give L and M: and points o and x to give J and K. Connect the twelve points on the circle to produce the dodecagon AMDKFHBLEJCG.

How to draw a thirteen-sided figure

Figure 10.7. Step 1
Draw a circle with centre O. Draw the diameter AB, and then two further circles with the same radius as the first circle, centred on A and B. These cross the first circle at C, D, E and F — which is a √3 rectangle, intersecting the horizontal at G. Set your compass to a radius of GE.

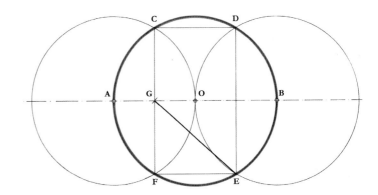

Figure 10.8. Step 2
GE is the length of a chord between first and fourth points (here numbered 1 and 2) on a thirteen-sided figure to an accuracy of 99.75%. To create a thirteen-pointed star, draw a series of arcs of radius GE, starting at point 1, and at each new point struck on the circumference of the circle. Thus our compass will be placed in turn on points 1, 2, 3, 4 and so on until, with the compass on point 13, our arc will run from point 12 back to our starting point at position 1. Draw lightly at first as you may need to slightly adjust the compass radius to achieve 100% accuracy. (Construction by Daniel Docherty.)

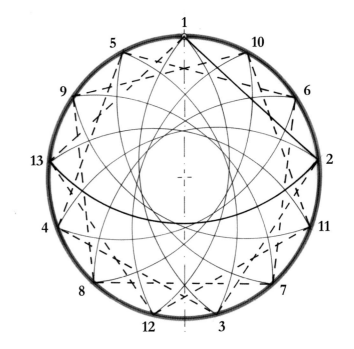

10. TEN, ELEVEN, TWELVE AND THIRTEEN-SIDED FIGURES

48

Chartres labyrinth

Left Drawing of the Chartres Labyrinth by the author.*

---

* See Critchlow, Vaughan Lee and Carroll (1975).

Right Detail of the thirteen-point centre of the Labyrinth. ▶

Below Recreation of the Chartres Labyrinth at New Harmony, Indiana, USA, by Jane Owen and Kent Schuette. ▼

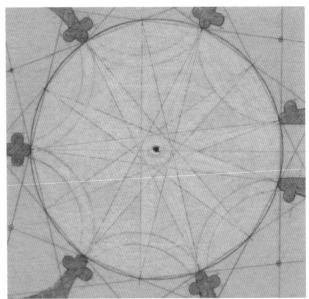

11. Dividing a Line

To divide a line into two

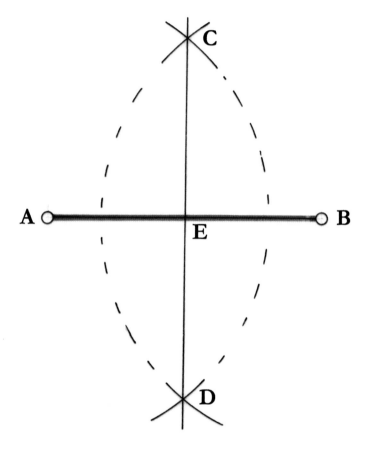

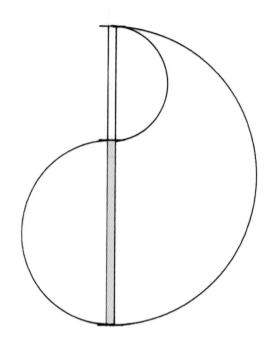

Figure 11.1.
To divide a line AB, or any section of a line, into two proceed as follows. With the compass on point A and radius greater than half the length of AB, draw an arc from above to below the line. With the same radius, and compass point on B, draw an arc to cross the first arc in two places — points C and D. Draw a line to connect C and D, crossing AB at point E, the mid-point of both lines AB and CD.

This method can be combined with the methods for dividing a line into three, five, seven and nine to divide a line into six, ten, and so on.

To divide a line into any number with a set-square

The following set-square method for dividing a line into any number of divisions will be familiar to architects and technicians trained on a drawing board.

Figure 11.2.
From one end of the line AB to be divided, draw a line AC at any convenient angle. Mark off the required number of divisions (in this case, three) along AC with a ruler, using the same measurement between each point. Set your set-square to connect the last measured off point F to the other end of the line to be divided, point B. With the set-square at the same angle, draw lines through the other measured off points D and E to cut AB at new points G and H. G and H divide AB exactly into three.

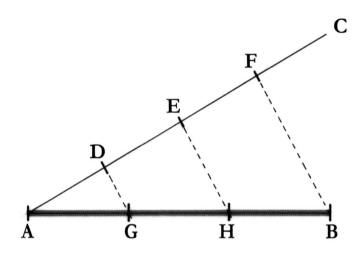

To divide a line into three

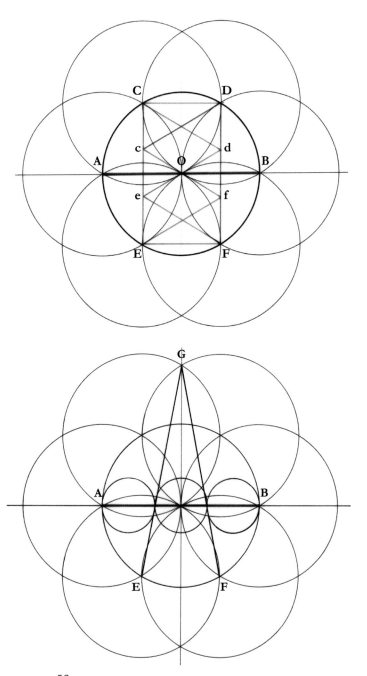

Figure 11.3.
The following constructions use the same base drawing. First draw the line AB which is to be divided. Find the centre point O of this line using the method described in Figure 11.1, and draw a circle with radius OA. With the same radius, draw two circles centred on points A and B. Then complete the *Flower of Creation* (see Figure 6.2) by drawing four further circles centred on the points C, D, E and F (where the previous circles crossed the first circle). The √3 rectangle and its 30 degree divisions (drawn in blue) can be drawn by first drawing lines CD, DF, FE, EC. Then draw a line from C towards B giving point d; A towards F giving point e; E towards B giving point f; and D towards A giving point c. Finally connect d to e through O; and c to f through O. We now have our base drawing for the following four constructions.

Figure 11.4.
AB can be divided into three by finding point G where circles C and D intersect, and then connecting points E and F to G as shown. AB is divided into three at the points where EG and FG intersect AB. You can draw the three circles shown above by placing your compass in the centre (point O) and setting a radius to the points on AB just found. Then with the same radius draw circles to each side, tangenting the first circle at points A and B. (Construction by Hana Hijazi.)

11. DIVIDING A LINE

To divide a line into five

Figure 11.5.
Find point H (crossing point of lines cD and dC), and connect E and F to H. AB is divided into a fifth at the points where EH and FH intersect AB. Use this fifth measure to fully divide the line into five. (Construction by Hana Hijazi.)

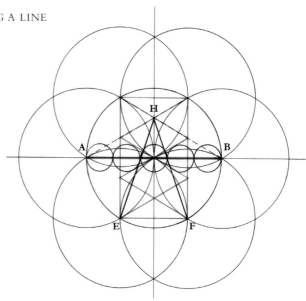

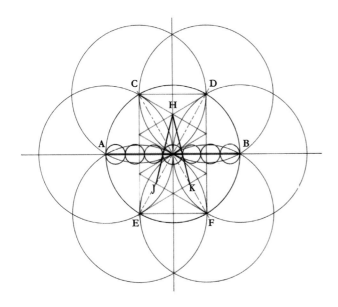

To divide a line into seven

Figure 11.6.
Find points J and K by drawing the diagonals ED and FC as shown. Connect J and K to H. AB is divided into a seventh at the points where JH and KH intersect AB. Use this seventh measure to fully divide the line into seven. (Construction by Hana Hijazi.)

To divide a line into nine

Figure 11.7.
In the hexagon within the √3 rectangle CDFE, construct the six-pointed star (hexagram) by joining points H, d, f, L, e and c. Connect new points M and N on the hexagram to H. AB is divided into a ninth at the points where MH and NH intersect AB. Use this ninth measure to fully divide the line into nine. (Construction by Hana Hijazi.)

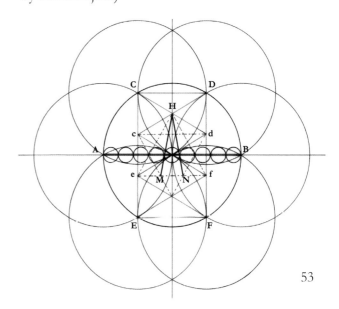

To divide a line into the golden section* (Φ)

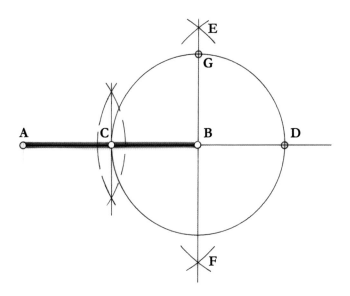

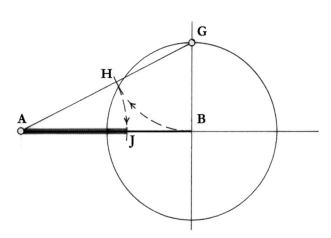

Figure 11.8. Step 1
AB is the line to be divided. First find the midpoint of AB by drawing arcs from A and B, with radius greater than half the distance between the two points, and connecting the intersections above and below the line to cut AB in half at C. With compass point at B and radius BC draw a circle. Extend AB to cut this circle at D. Draw the perpendicular through B by drawing arcs from C and D, and connecting their intersections (E and F) to find point G on the circle. GB has the same value as CB, which is AB/2.

Figure 11.9. Step 2
Draw in line AG. With compass on G and radius GB, draw an arc from B to cut AG at H. With compass on A and radius AH, draw an arc from H down to cut AB at J. AB:AJ has the ratio Φ:1.**

---

* One of the best expositions of the meaning of this most significant of ratios can be found in *Parabola* Vol. XVI, No.4. See also Olsen 2006.

** Φ is the solution to the quadratic equation $x^2 - x - 1 = 0$. The value $(\sqrt{5} + 1)/2$ is commonly given for Φ.
In triangle ABG above, if AB=1 then GB=1/2. By Pythagoras ($x^2 + y^2 = z^2$) or in this case $AB^2 + GB^2 = AG^2$ we find $AG = \sqrt{5}/2$.
So $AH = \sqrt{5}/2 - 1/2$, or $\sqrt{(5-1)}/2$
which is also value of AJ.
Φ has extraordinary numerical properties:
Φ = 1.618033989 ... Φ² = 2.618033989 ...
1/Φ = 0.618033989, which is also the value of $\sqrt{(5-1)}/2$, and also (Φ−1).

11. DIVIDING A LINE

To find a golden mean proportion along a line

Figure 11.10. Step 1
Draw a line AB. Draw two circles on A and then B with radius AB. Draw the vertical axis by connecting the two points where the circles intersect (C and D). Draw a line from C through B to meet circle B at E. Similarly, draw a line from C through A to meet circle A at F. Join E to F, giving us the equilateral triangle CEF.

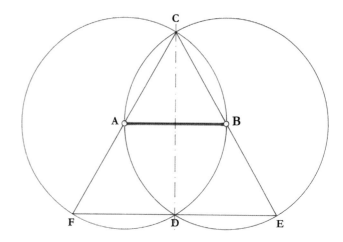

Figure 11.11. Step 2
Draw a line from E to A, and from F to B. These lines will cross the vertical axis at O, which is the centre of the triangle. Draw a circle centred on O with radius OC. Extend AB to meet this circle at points G and H. GA and BH are in golden mean ratio to AB. That is, GA:AB:BH=1:Φ:1.

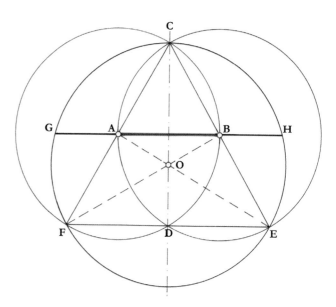

55

12. Miscellany

Squaring the circle (circling the square)

Drawn 'solutions' to this philosophical, alchemical and psychological paradox are numerous, and involve constructing a circle whose circumference has the same length as the perimeter of a given square. Given below are one measured and one geometric solution — both 99.9% accurate.

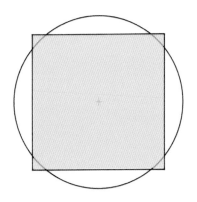

Method 1*

Figure 12.1.
Start by drawing a line AB measuring 11 measures of a convenient unit. Draw a square (see Figure 4.4) measuring 11 × 11 units, and find the centre by drawing the diagonals CB and AD. Now construct two 3:4:5 triangles on AB — (this can be done by swinging arcs with radius of 5 units from A and B, and then arcs measuring 3 units from points 4 units along from each end (A and B). The arcs will give the upper corners of a 3 × 3 square. Find the centre (point E) of this square by drawing the diagonals. A circle centred on O with radius OE will have the same circumference as the 11 × 11 square.**

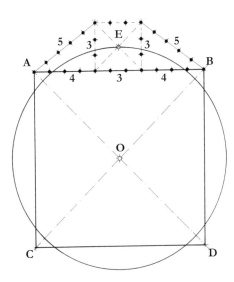

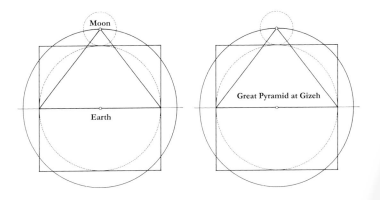

---

* As discussed in Chapter 6 of John Michell (1973) this construction not only gives the relative sizes of the earth and the Moon, but also the fundamental proportions of the Great Pyramid.

** Mathematically this construction is 100% if one takes the traditional equivalent for π of 22/7.

56

Method 2*

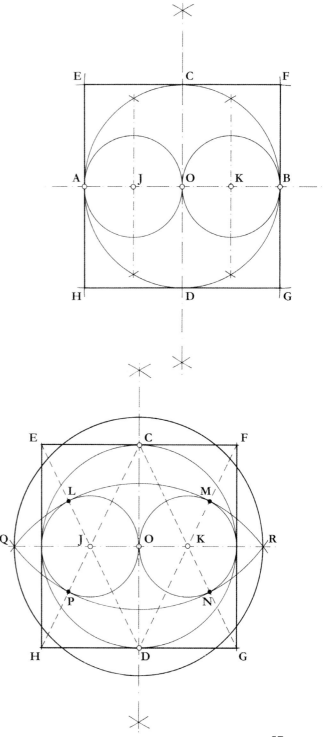

Figure 12.2. Step 1
First draw a circle centred on O, and draw a horizontal axis passing through A and B on the circle. Draw arcs from A and B to intersect above and below the circle to find the vertical axis, which crosses the circle at C and D. Construct the square EFGH about the circle by drawing arcs from A, C, B and D with radius OA. Now bisect AO and OB to find J and K, and draw circles centred on J and K with radius OK.

Figure 12.3. Step 2
This diagram shows another method for finding points J and K by drawing diagonals in the half squares ECDH and CFGD: these intersect the horizontal at J and K. These diagonals give us points L, M, N and P. Now set the compass on D and draw an arc to tangent the small circles at L and M; this will cross the horizontal axis at Q and R. A similar arc from point C will tangent the small circles on the lower side at P and N, and also meet the horizontal axis at Q and R. Now draw a circle with centre at O and radius OQ: this circle has the same circumference as the square EFGH.

---

* Given in Robert Lawlor *Sacred Geometry* (1982). See also R. Schwaller de Lubicz, *Temple of Man*.

12. MISCELLANY

A question of coincidence

How closely do the points A, B and A', B' on the diameter in the following two constructions coincide?

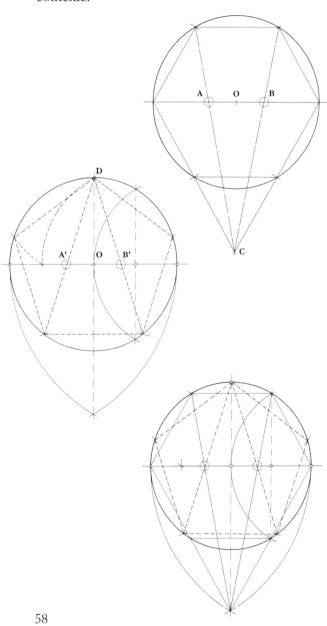

The first comes from a hexagonal construction, the second from a pentagonal construction:

Figure 12.4.
Point C is found by extending the two lower sides of the hexagon until they meet. Lines drawn from C to the two upper points of the hexagon will divide the diameter exactly into three.

Figure 12.5.
This construction of the pentagon can be found on page 20 (Figures 5.5 and 5.6). The golden triangle constructed on the base line will cross the diameter of the circle at points A' and B'.

We can find how closely the two sets of points A B and A' B' coincide by considering triangle OA'D in Figure 12.5. We know the angle of the triangle within a pentagon is 36°, so angle A'DO is 18°. If we let the diameter of the circle = d, so that the radius (which is OD)=d/2, then:
A'O=tan 18° × d/2=0.325 × d/2. So A'B' which is twice the length of A'O=0.325 × d. But AB, from Figure 12.4, is one third of the diameter. So the accuracy of the coincidence is 0.325/0.333 ... which is 97.5%.

Figure 12.6.
Both constructions shown together on the same diagram.

Polygon stars

There are no stars within the triangle or square:

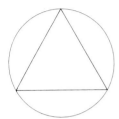 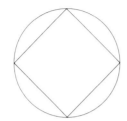

There is one star within the pentagon and hexagon:

 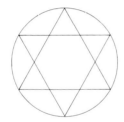

There are two stars within the heptagon and octagon:

 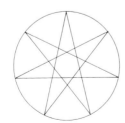 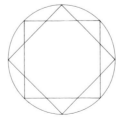 

There are three stars within the enneagon and decagon:

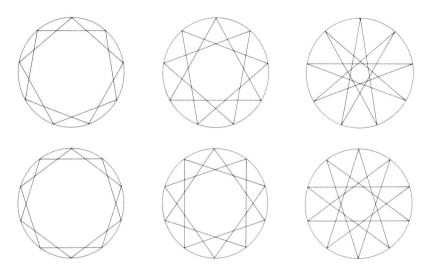

There are four stars within the endecagon and dodecagon:

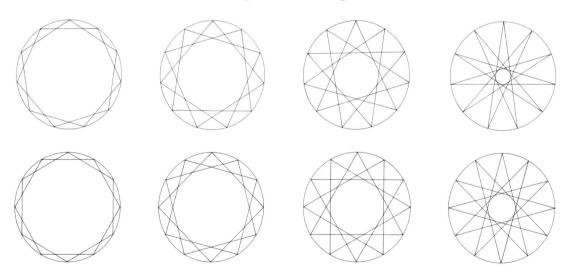

and so on ...

12. MISCELLANY

A progression of threes

Figure 12.7.
Having drawn an equilateral triangle within a circle (Figures 3.1 and 3.2), draw a circle inside the triangle to tangent the three sides. The exact points of tangent can be found by drawing lines through the centre from each corner. At the tangent points, a new smaller equilateral triangle can be drawn. You can then repeat the process as many times as you wish (and the scale of your drawing permits). The drawing above shows a progression of four triangles, each a quarter of the previous triangle. (See Critchlow 1969, p.59.)

A progression of fours

Figure 12.8.
To draw a progression of squares, first construct a (dynamic) square within a circle and draw on the diagonals (see Figures 4.1 and 4.2). Connect the points where the diagonals cross the first square to find the second (static) square, then the points where the vertical and horizontal axes cross this square to draw the third. Proceeding in this manner, a progression of diminishing squares, alternately dynamic and static (that is, rotating by 45° each time) is drawn.* (See Critchlow 1969, p.59, also pp.11, 91.)

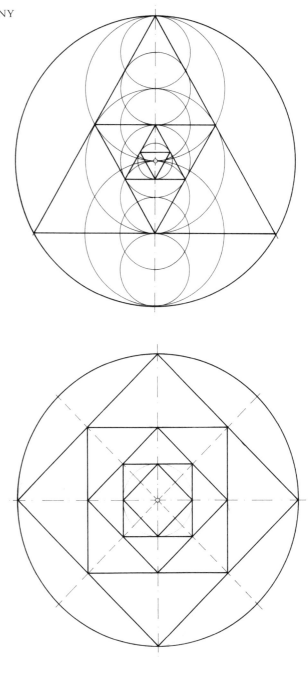

* The relationship of one square to the next is in the ratio of √2:1.

12. MISCELLANY

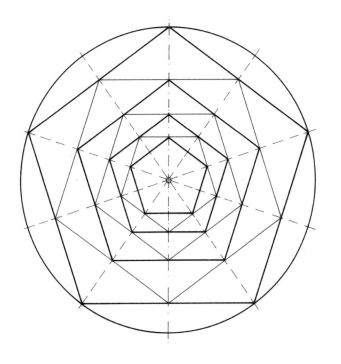

A progression of fives

Figure 12.9.
By drawing a pentagon (e.g. Figures 5.1–5.4), adding the diagonals, and following an alternating method as for the progression of squares (Figure 12.8), a progression of pentagons can be drawn.

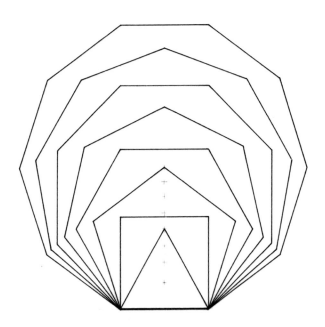

A progression of polygons

Figure 12.10.
Although this book has largely dealt with the basic polygons separately and as simple shapes, beautiful and profound proportions can be found not only within each polygon, but also in their combination. The whole subject of proportion must be left to another volume. For now, the diagram on the left stands as a reminder that the polygons can be seen as members of a family.

12. MISCELLANY

A marriage of polygons

Figure 12.11.
This diagram, combining the hexagon, pentagon and square, has traditionally stood for the relationship between 'Heaven,' 'Man'* and 'Earth.'** It is instructive to note that whereas 'Heaven' only just touches 'Earth,' 'Man' occupies equally an area of both. Where in each case the overlapping 'circle of Man' creates a vesica, we could see this as representing respectively the field of creation of his 'divine nature' and the field of creation of his 'human nature' — both together making a complete human being.

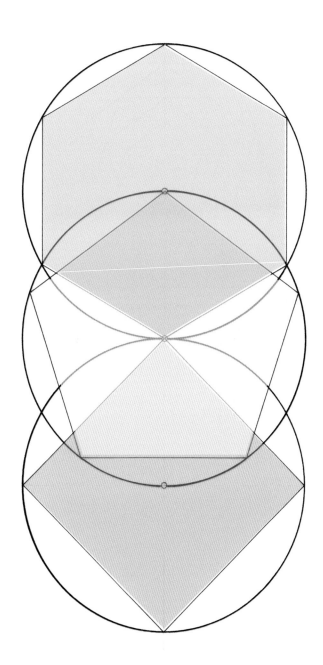

---

* 'Man' here meaning Mankind rather than a male human, from the sanskrit 'manas' indicating 'conscious animal'.

** Another set of labels for this fundamental cosmic creative progression is; 'Unity,' 'Consciousness' and 'Matter.' Yet another, in reverse order: *sat-chit-ananda*.

Appendix

A note on accuracy

What does it mean to be *accurate*? How *precise* is it possible to be? How *exact* is it necessary to be?

The answer to these questions will depend, of course, on the situation. For example, a builder will construct a brick wall to a certain 'tolerance' (or accuracy), which means that his wall will comply with the appropriate technical standards if it is vertical to within a stipulated measurement; and in practice, a three storey house could have its walls as much as a couple of inches out of plumb and still be technically 'vertical'. There is more latitude in the measurement of ingredients for cooking than for chemical experiments in the laboratory: and the levels of precision required in the glass of astronomical telescopes or computer components are of quite different orders again. None, however, are absolutely precise.

Mathematics generally deals with exact quantity (number), with the notable exception of the 'transcendental' values such as $\sqrt{3}$, $\sqrt{2}$ and Φ found in geometry, which can be easily drawn but are incapable of numerical resolution.

Philosophically it could be argued that absolute precision is a quality of perfection, and thus not of 'this' world. In this world of material creation, the Universe, a certain inexactitude (what we might call 'the wobble factor') would appear to be necessary, and part of the explanation of how things come into being — for in absolute perfection there is no need for movement.

The calculations that follow are included to substantiate the accuracy percentages stated for the various constructions in this book. They may be of interest to those with a mathematical turn of mind.

It is worth noting that the slight imperfections of constructional geometry do not compromise its practical application in art and craft, where 'imprecision' is accommodated by the experienced eye and hand which naturally make the necessary adjustments. Indeed, it is perhaps by learning those necessary subtle adjustments that a deeper understanding of the art or craft is achieved.

APPENDIX

Pentagon in a circle

Method 1

Page 18

Let the radius of large circle O = 2 = OL
so radius of smaller circles G and H = 1 = OH, HT.

In triangle LOH,
LH = √(OH² + OL²) = √(4 + 1) = √5
and LH = LT + HT
so LT = √5 − 1 = 1.23606798 = LN.

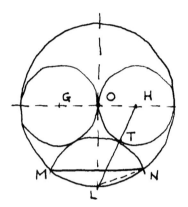

But in a regular Pentagon
again with radius = 2 = ON
LN/2 = 2 × sin18°
∴ LN = 4 × sin18° = 1.23606798
which is the same value as above
so our construction is 100% accurate.

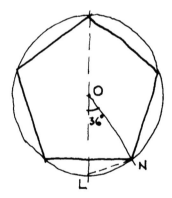

Pentagon in a circle

Method 2

Page 20

Let the radius of the circle = 1 = OF
so OE = 0.5

In triangle FOE
EF = √(1 + 0.25) = √5/2 = EG.

In triangle FOG
OG = EG − OE = √5/2 − ½ = (√5 − 1)/2
so FG = √(OG² + OF²)
    = √([√5 − 1)/2]² + 1²)
    = 1.175570505 = FH.

But in a regular Pentagon
half side length x = cos54°
so whole side length = 2 × cos54°
                    = 1.175570505
which is the same value as above
so our construction is 100% accurate.

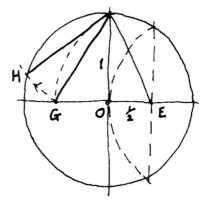

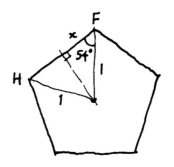

APPENDIX

Pentagon in a vesica

Page 21

Let AB = 1, AO = 0.5
so Vesica constructed on points A and B has height of √3
∴ CO = √3/2.

Circle centred on O with radius OB = ½
is then drawn centred on C
intersecting vertical axis at E
so EO = CO − CE
    = √3/2 − 0.5
    = 0.3660254

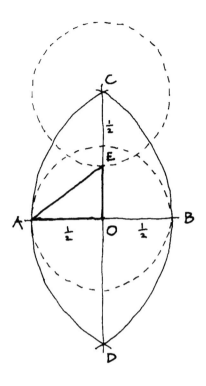

But if AE is the side of a Pentagon
in triangle AOE we would have angle AEO = 54°
giving tan54° = AO/EO
so EO = AO/ tan54°
    = 1/2 × tan54°
    = 0.3632713

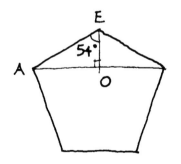

Comparing the two values, our construction is 99.25% accurate.

Pentagon from a hexagon

Page 22

Let radius of circle = 1 = CD, EF
and also DH (arc from C centred on D).

In rectangle CDFE
DF = √3
DH = 1
so HF = √3 −1

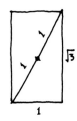

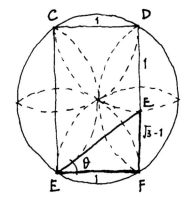

In triangle HEF
if θ is ∠HEF
tanθ = HF/EF = (√3 −1)/1
so θ = 36.206°

But in a regular Pentagon the shoulder angle = 36°
so our construction is 36/36.206 = 99.43% accurate.

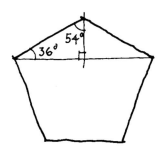

Pentagon on a line

Method 1

Page 23

Let OB = 1 = AO
so AB = 2 = BE.

In triangle OBE
$OE^2 = OB^2 + BE^2$
$\quad\quad = 1 + 4 = 5$
so $OE = \sqrt{5} = OH$.

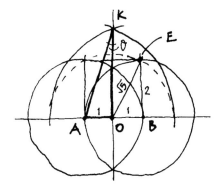

In triangle AOK
AK = AH = AO + OH = $1 + \sqrt{5}$
and if θ is $\angle AKO$
$\sin\theta = 1/(1 + \sqrt{5})$
$\therefore \theta = 18°$
which is precisely the angle in a regular Pentagon.

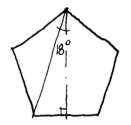

APPENDIX

Pentagon on a line

Method 2

Page 25

Let AB = CB = BH = FC = 1
so AX = XB = 0.5

In triangle CXB
XC = 1 × cos30° = 0.8660254

FX = FC − XC
   = 1 − 0.8660254 = 0.1339746 = YX
(45° triangle)
and YB = YX + XB
   = 0.1339746 + 0.5
   = 0.6339746

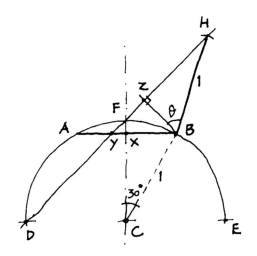

In triangle YBZ
BZ = YB × sin45° = 0.4482877

In triangle HBZ
if θ is ∠HBZ
$\cos\theta$ = BZ = 0.4482877/1
∴ θ = 63.366120°

So angle ABH = 45 + 63.366120° = 108.366120°

But internal angle of a regular Pentagon is 108°
so our construction is 108/108.36612 = 99.66%
accurate.

70

Pentagon on a line

Method 2 — Improved

Page 26

We have from previous page $XB = 0.5$, $XC = 0.8660254$
and $FC = DC = 1$

In triangle DCZ
$DC^2 = CZ^2 + DZ^2$
but $CZ = DZ$, and $DC = 1$
so $1 = 2 \times CZ^2$
$\therefore CZ = \sqrt{0.5} = 0.70710678 = CJ$

In triangle BKX
$XK = XC + CK = 0.8660254 + 0.70710678 = 1.57313218$
and if θ is $\angle BKX$
$\tan\theta = XB/XJ$
$ = 0.5/1.57313218$
$ = 0.31783725$
$\therefore \theta = 17.6322°$

But true Pentagonal angle here is $18°$
so our construction is only 97.96% accurate.

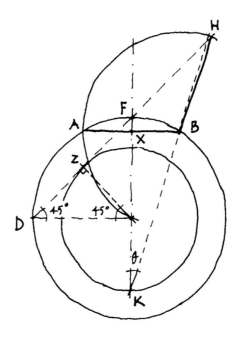

Pentagon on a line

Method 3

Page 27

Let radius of circles = 2 = AB
so AO = 1

In triangle AOG
if θ is ∠AGO
tanθ = 1/3
so θ = 18.43495°

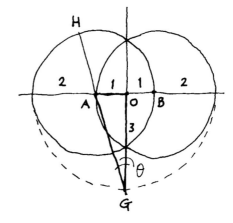

But angle made by regular Pentagon at this point is 18° so our construction is 18/18.43495 = 97.64% accurate.

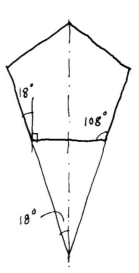

Heptagon in a circle

Method 1

Page 30

Let radius of circle = 1 = OC, OK, EC, CF
so as EFJ is an equilateral triangle, EJ = EF = 2
and $\angle JEC = 60°$

In triangle JEC
$\tan 60° = CJ/1$
so $CJ = \tan 60° = 1.732050808$
but $CO = 1$
$\therefore OJ = CJ - CO = 1.732050808 - 1 = 0.732050808$

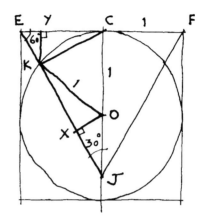

In triangle OJX
$XJ = OJ \times \cos 30° = 0.633974596$
and $OX = OJ \times \sin 30° = 0.366025404$

In triangle KOX
$KX^2 + OX^2 = OK^2$
so $KX = \sqrt{(1 - OX^2)} = 0.93060486$

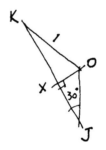

As $EJ = 2 = EK + KX + XJ$
so $EK = 2 - (KX + XJ) = 0.43542055$

In triangle YEK
YK = EK × sin60° = 0.37708525
and EY = EK × cos60° = 0.21771027
But EC = 1
so YC = 1 − EY = 0.78228973
In triangle YCK
if θ is ∠YCK
tanθ = YK/YC
so θ = 25.735°

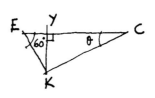

But θ is half the internal angle of a regular
Heptagon, which is 360/14 = 25.714°
so our construction is 25.714/25.735
= 99.9% accurate.

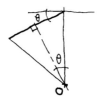

Heptagon in a circle

Method 2

Page 32

Let radius of circle = 1 = OA = AC
OAC is an equilateral triangle.

In triangle AOX (AX bisects ∠OAC)
AX = sin60° = 0.8660254 = AE

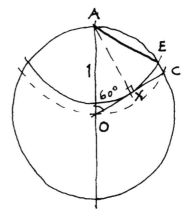

But in a regular Heptagon
length of side = 2 × sin360/14 = 0.86776748
so our construction is 0.8660254/0.86776748
= 99.79% accurate.

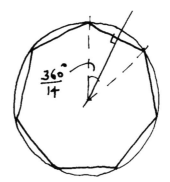

Heptagon in a circle

Method 3

Page 33

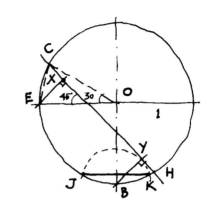

Let radius of circle = 1 = OE, OB, OC

In triangle EOC
EC = 2 × sin15° = 0.51763809

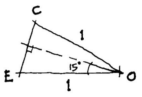

In triangle CEX
EX = EC × cos30° = 0.4482877 = BY = BK

In triangle ZOK
OK = OB = 1, ZK = BK/2
and if θ = ∠ZOK
tanθ = BK/2
so θ = 12.633686°

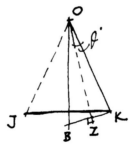

Now ∠JOK = 4 × ∠θ = 50.534743°
but the internal angle of a regular Heptagon is
360/7 = 51.428571°
so our construction is 50.534743/51.428571
= 98.26% accurate.

How to draw a heptagon

Method 4

Page 34

Let the unit of measurement used = 1
(so OA = 5)

In triangle DOK
OD = 4 (radius of circle), and OK = 5/2 (half OA)
so $DK^2 = OD^2 - OK^2$
$ = 4^2 - (5/2)^2$
$ = 39/4$

In triangle DKB
KB = 3/2
so $DB^2 = DK^2 + KB^2$
$ = 39/4 + 9/4$
$ = 48/4 = 12$
$\therefore DB = \sqrt{12}$

In triangle DOB
if $\angle DOB = \theta$
$\sin\theta/2 = DB/2 \div DO = \sqrt{12}/2 \div 4$
so $\theta/2 = 25.6589063°$
and $\therefore \theta = 51.3178126°$

But internal angle of a Heptagon
= 360/7 = 51.4285714°
so our construction is 51.3178126/51.4285714
= 99.78% accurate.

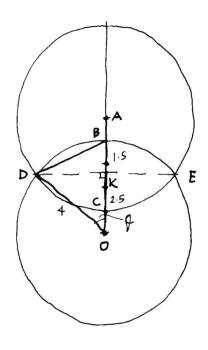

APPENDIX

Heptagon in a vesica

Page 35

Let AB = 1 = AC, BC
so AO = OB = ½
and EO = EB = ¼

In triangle COB
$CB^2 = CO^2 + OB^2$
so CO = √(1 − ¼) = √3/2
but CF = EO = ¼
so OF = √3/2 − ¼ = 0.6160254

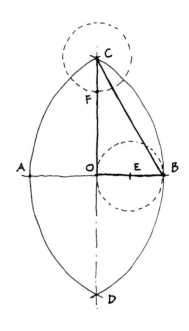

For a regular Heptagon
in triangle AXO
let ∠AXO = θ
Now as ∠AXB = 3 × 360/7,
θ = ½ × 3 × 360/7 = 77.142857°
Thus OX = 1/(2 × tanθ) = 0.1141217
and AX = 1/(2 × sinθ) = 0.512858 = XF

This gives us OX + XF = 0.1141217 + 0.512858
= 0.626980 = OF
But our value for OF above was 0.6160254
so our construction is 0.6160254/0.626980
= 98.25% accurate.

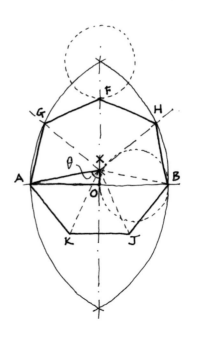

Enneagon in a circle

Page 40

Let radius of circle OA = OB = OC = 1
so CD = 2

As CE = CA = CB = √2 (diagonal of 45° triangle of side = 1)
ED = 2 − √2 (CD−CE)
so EH = (2 − √2)/2
∴ CH = √2 + (2 − √2)/2

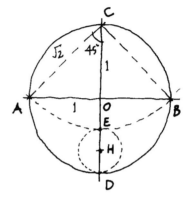

In triangle HCX
HX = EH
and if ∠HCX = θ
sinθ = HX/CH
so θ = 9.879282°

But in a regular Enneagon
tanθ = tan20°/2
which gives θ = 10.314105°
so our construction is only 9.879282/10.314105
= 95.78% accurate.

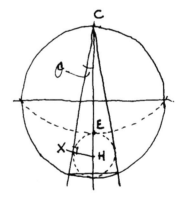

APPENDIX

Enneagon in a vesica

Page 41

Let $AB = CB = 2$, so $AO = OB = 1$, and $OK = KB = 0.5$
Also $CO = \sqrt{3}$ (vertical of equilateral triangle of side = 2).

In triangle COK
$CK^2 = CO^2 + OK^2$
$ = 3 + 0.25$
so $CK = 1.8027756$
but $CK = CX + XK$
and $XK = OK = 0.5$
$\therefore CX = 1.8027756 - 0.5$
$ = 1.3027756$

In triangle COB
$CB = 2$, and $CM = CX = 1.3027756$ from above
$\therefore MB = 2 - 1.3027756 = 0.69722436$

If MB is one side of a regular Enneagon
in triangle OSB
$\angle OSB = 80°$ (twice the internal angle of $360/9°$)
so $SB = OB/\sin 80° = 1/\sin 80°$
$ = 1.0154266$
And in triangle BSM
where $\angle BSM = 40°$
$MB/2 = \sin 20° \times SB$
giving $MB = 0.6945927$

So our construction is $0.6945927/0.69722436$
$= 99.62\%$ accurate.

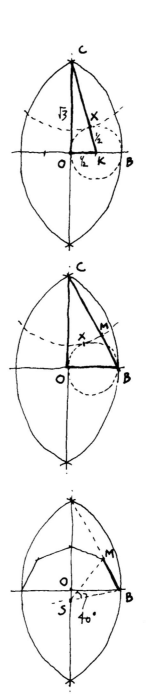

80

APPENDIX

Eleven-sided figure

Page 45

Let the radius of the large circle = R, and the radius of the four intermediate sized circles = r
so there are seven radii of 'r' in the diameter of the large circle 2R
∴ 2R = 7r

In triangle LOW
where LO = R, and LW = r/2
let ∠LOW = θ
so sin θ = r/2R
but as 2R = 7r
sin θ = 1/7
so θ = 8.213210°
and ∴ ∠LOM = 4θ = 32.852843°

But the internal angle of an eleven-sided figure = 360/11° = 32.727272̇°
So our construction is 32.727272/32.852843 = 99.62% accurate.

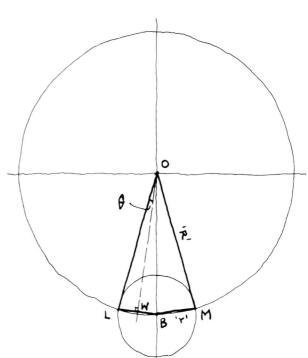

Thirteen-sided figure

Page 47

Let the radius of the circle = 1 = EF

In a regular 13-sided figure, the length of a chord $2x$ from any point to another point missing two can be found from:
$x = \sin(1.5 \times 360/13)° = 0.66312266$
giving $2x = 1.3262453$

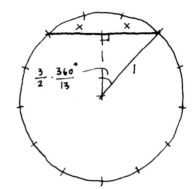

In triangle GEF in our construction
GF = $\sqrt{3}/2$ (height of an equilateral triangle of side = 1)
so $GE^2 = EF^2 + GF^2$
∴ GE = $\sqrt{(1 + 0.75)}$
     = 1.3228757

Comparing the two values, our construction is
1.3228757/1.3262453 = 99.75% accurate.

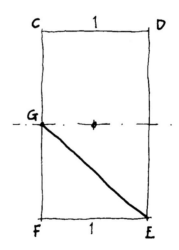

APPENDIX

Squaring the circle

Method 1

Page 56

Each side of the square measures 11
so the perimeter of the square measures 44

The circumference of the circle is given by $2\pi r$
and the radius of the circle is 7
so the circumference is $14 \times \pi = 43.9823$
which is accurate to $43.9823/44 = 99.96\%$

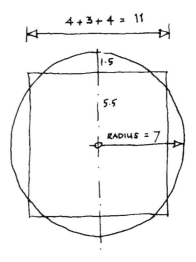

Squaring the circle

Method 2

Page 57

Let the radius of the circle OD = OA = 1
which is also half the side length of our square
whose perimeter is thus 8

In triangle JOD
OJ = OA/2 = ½
so DJ² = OJ² + OD²
    = ¼ + 1
∴ DJ = √5/2

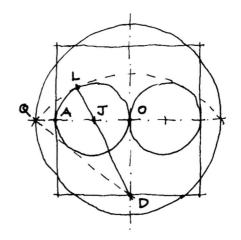

In triangle QOD
QD = DJ + JL = √5/2 + ½ = (√5 + 1)/2
but JL = OJ
so QO² = QD² − OD²
∴ QO = 1.2720197
which is the radius of our constructed circle
the circumference of which is therefore 2πQO
 = 7.992335

But the perimeter of the square is 8
so our construction is 7.992335/8 = 99.9% accurate.

Observing that the value for QD is also the formula
for Φ
we could write the above equation for QO as
QO² = Φ² − 1 = Φ
so QO = √Φ = 1.2720197

Afterword: Explorers in Geometry

Kairos is an educational charity and a non-sectarian organization specifically founded to promote the recovery of traditional (perennial) values in the Arts and Sciences.

Kairos means both proportion and timely, which summarizes the four studies of the Quadrivium: Arithmetic, Geometry, Music and Astronomy. Kairos is pleased to receive any material that pertains to these subjects — from any quarter, or any tradition — particularly where this increases an understanding of the relationships between arithmetic, geometry, music (harmony) and astronomy (cosmology) which are the universal languages of humankind.

Many have contributed over the years to our current explorations in geometry, through their writings, or through classes and workshops both in the USA and in the UK, and all deserve to be acknowledged:

John Michell, Paul Marchant, Michael Ben Eli, Jonathan Horning, Steve Bass, Peter Gilbert, Julian Carlyon, Farooq Hussain, David Green, Carl Kowsky, David Tasker, David Marks, Julia Barfield, Llewellyn Vaughan-Lee, Jane Carroll, Richard Waddington, Robert Lawlor, Rachel Fletcher, Robert Meurant, Scott Olsen, Lance Harding, John Martineau, Ramiz Sabbagh, David Feurstein, Richard Chaffer, Nic Cope, Adam Tetlow, Richard Henry, Tom Bree, Zara Hussain, Lisa de Long, David Barnes, John Lloyd, Ana Maria Giraldo, Daniel Docherty, Hana Hijazi, and my teacher Sergei Kadleigh.

It would be quite wrong not to mention the great masters who, through the centuries, transmitted the impetus and wisdom of geometry in the first place. Added to these, we should mention all those scholars and geometricians who are dedicated to recovering the science, wonder and awe of the great church and cathedral builders of the past. The present revival of interest in the intrinsic (mostly esoteric) geometry of the great sacred monuments of the world is based on our modern sense of loss of essential learning. In its place, we find overloads of information, for which we have to thank the modern computer. The real relevance of the classical Quadrivium, the foundation of the European cathedral schools tradition, is returning with a vital seriousness. This volume by Jon Allen will aid the revival and become a blessing to those who read and practise its work, in ways they may not have imagined.

Keith Critchlow
March 2007

Further Reading

Critchlow, K.B. (1969) *Order in Space,* Thames & Hudson, London.

—, (1976) *Islamic Patterns,* Thames & Hudson, London.

—, (1979) *Time Stands Still,* Gordon Fraser Gallery, London (reissued 2007, Floris Books, Edinburgh).

—, Vaughan Lee, L., Carroll, J. (1975) *Chartres Labyrinth,* RILKO publication. Reprinted by Kairos 2002.

Ghyka, M. (1952) *A Practical Handbook of Geometrical Composition and Design,* Alec Tiranti, London.

Lawlor, R. (1982) *Sacred Geometry: philosophy and practice,* Thames & Hudson, London.

Lundy, M. (2000) *Sacred Geometry,* Wooden Books, Wales.

Michell, J. (1973) *City of Revelation,* Sphere Books, London.

Martineau, J. (1995) *A Book of Coincidence,* Wooden Books, Wales.

Olsen, S. (2006) *The Golden Section,* Wooden Books, Wales.

Parabola, Vol.XVI, No.4, November 1991. *The Golden Proportion: a conversation between Richard Temple and Keith Critchlow.*

Schwaller de Lubicz, R.A. (1977) *The Temple in Man,* Inner Traditions, Rochester, Vermont, USA. Trans. R. & D. Lawlor.

About the Author

Jon Allen was a practising architect for more than twenty years. He worked closely with Keith Critchlow, a world authority on geometry, and had a specialist interest in the application of geometry to architectural design.

Since retiring from his architecture practice, Jon has worked as a Geometry Consultant. He lives in London. He is happy to receive feedback on this book, and may be contacted via his website:

www.jonallenarchitect.co.uk

Making Geometry

Exploring Three-Dimensional Forms

Jon Allen

Following on from his successful *Drawing Geometry*, Jon Allen explores the creation of the many-sided three-dimensional forms known as the Platonic and Archimedean solids.

Many professionals find they need to be able to build three-dimensional shapes accurately, and understand the principles behind them. This unique book demonstrates how to make models of all the Platonic and Archimedean solids, as well as several other polyhedra and stellated forms. It provides step-by-step instructions for constructing the three-dimensional forms, as well as showing how to draw out accurately the geometry of the paperfold nets.

Beginners and experienced artists and designers alike will find this book a source of practical guidance, as well as delight and inspiration which will amply repay the careful attention needed to construct the models.

www.florisbooks.co.uk